100
Numerical Games

Also by Pierre Berloquin

100 GEOMETRIC GAMES

PIERRE BERLOQUIN

100
Numerical
Games

Foreword by Martin Gardner
Drawings by Denis Dugas

CHARLES SCRIBNER'S SONS · NEW YORK

1 3 5 7 9 11 13 15 17 19 H/C 20 18 16 14 12 10 8 6 4 2

Printed in the United States of America
Library of Congress Catalog Card Number 76-21996
ISBN 0-684-14668-1

Contents

Foreword

Pierre Berloquin, who put together this stimulating and delightful collection of mind benders, is a clever young Frenchman who was born in 1939 in Tours and graduated in 1962 from the École Nationale Supérieure des Mines in Paris. His training as an operations research engineer gave him an excellent background in mathematics and logical thinking.

But Berloquin was more interested in writing than in working on operations research problems. After two years with a Paris advertising agency, he decided to try his luck at freelance writing and this is how he has earned his living since. In 1964 he began his popular column on "Games and Paradoxes" in the magazine *Science et Vie* (Science and Life). Another column, "From a Logical Point of View," appears twice monthly in *The World of Science*, a supplement of the Paris newspaper *Le Monde*. Occasionally he contributes to other French magazines. One of his favorite avocations is leading groups of "*créativité*," a French cocktail of brainstorming, synectics, and encounter therapy, for the discovery of new ideas and the solution of problems—a logical extension of his interest in puzzles.

Berloquin's published books are *Le Livre des Jeux* (card and board games), *Le Jeu de Tarot* (tarot card game), *Testez Votre Intelligence* (intelligence tests), *100 Grandes Réussites* (solitaire games), *Un Souvenir d'Enfance d'Evariste Galois* (Memoir of the Childhood of Evariste Galois) and *100 Jeux de Cartes Classiques* (card games); he is coauthor of *Voulez-Vous Jouer avec Nous* (Come Play with Us) and *Le Livre des Divertissements* (party games).

This volume is Berloquin's own translation into English of

one of his four paperback collections of brainteasers which have been enormously popular in France and Italy since they were published in Paris in 1973. This one is concerned only with numerical puzzles. The other three contain geometric, logical, and alphabetical problems. Denis Dugas, the graphic artist who illustrated all four books, is one of the author's old friends.

The puzzles in this collection have been carefully selected or designed (many are original with the author or artist) so that none will be too difficult for the average reader who is not a mathematician to solve, and at the same time not be *too* easy. They are all crisply, clearly given, accurately answered at the back of the book, and great fun to work on whether you crack them or not.

At present, Berloquin is living in Neuilly, a Paris suburb, with his wife, Annie, and their two children.

Martin Gardner

PROBLEMS

Game 1

Timothy spent all his money in five stores. In each store, he spent $1 more than half of what he had when he came in.

How much did Timothy have when he entered the first store?

Game 2

How many ways can you read ACE off the diagram? You can move horizontally, vertically, or diagonally.

Game 3

Let us follow the hour hand and minute hand of a clock for 24 hours.

How many times do they form a right angle?

Game 4

$$2 \ 2 \ 2 \ 2 = 0$$
$$2 \ 2 \ 2 \ 2 = 1$$
$$2 \ 2 \ 2 \ 2 = 2$$
$$2 \ 2 \ 2 \ 2 = 3$$
$$2 \ 2 \ 2 \ 2 = 4$$
$$2 \ 2 \ 2 \ 2 = 5$$
$$2 \ 2 \ 2 \ 2 = 6$$
$$2 \ 2 \ 2 \ 2 = 10$$
$$2 \ 2 \ 2 \ 2 = 12$$

Add arithmetical symbols between the 2's to make every equation true. You may use plus, minus, times, and divide symbols, as well as parentheses and brackets for grouping.

Game 5

Two towns are linked by a railroad. Every hour on the hour a train leaves each town for the other town. The trains all go at the same speed and every trip from one town to the other takes 5 hours.

How many trains are met by one train during one trip?

Game 6

$$3\ 3\ 3\ 3\ =\ 3$$
$$3\ 3\ 3\ 3\ =\ 4$$
$$3\ 3\ 3\ 3\ =\ 5$$
$$3\ 3\ 3\ 3\ =\ 6$$
$$3\ 3\ 3\ 3\ =\ 7$$
$$3\ 3\ 3\ 3\ =\ 8$$
$$3\ 3\ 3\ 3\ =\ 9$$
$$3\ 3\ 3\ 3\ =\ 10$$

Add arithmetical symbols between the 3's to make every equation true.

Game 7

A clock strikes every hour—once at 1:00, twice at 2:00, and so on.

The clock takes 6 seconds to strike 5:00 and 12 seconds to strike 9:00. The time needed to strike 1:00 is negligible.

How long does the clock need for all its striking in 24 hours?

Game 8

$$4 \ 4 \ 4 \ 4 = \ 3$$
$$4 \ 4 \ 4 \ 4 = \ 6$$
$$4 \ 4 \ 4 \ 4 = \ 7$$
$$4 \ 4 \ 4 \ 4 = \ 8$$
$$4 \ 4 \ 4 \ 4 = 24$$
$$4 \ 4 \ 4 \ 4 = 28$$
$$4 \ 4 \ 4 \ 4 = 32$$
$$4 \ 4 \ 4 \ 4 = 48$$

Add arithmetical symbols between the 4's to make every equation true.

Game 9

Timothy goes to a fountain which delivers an unlimited amount of water. He brings two empty containers, one of 7 liters, the other of 11 liters.

How many operations does he need to fill one of the containers with exactly 6 liters of water? (In these container games, in each operation one container must be completely filled or completely emptied.)

Game 10

8	1	6
3	5	7
4	9	2

This is a magic square. On each horizontal, each vertical, and each of the two diagonals the numbers always add up to 15.

Can you complete the next square with the integers from 5 through 16 to make it magic? On each horizontal, each vertical, and each of the two diagonals the sum of the numbers is to be 34.

(Hint: The two unfilled central squares must add up to 29; the two unfilled corner squares must also add up to 29.)

1			
		2	
	3		
			4

Game 11

If 73 hens lay 73 dozen eggs in 73 days and if 37 hens eat 37 kilograms of wheat in 37 days, what weight of wheat corresponds to 1 dozen eggs?

Game 12

```
        P
      P O P
    P O P O P
      P O P
        P
```

How many ways can you read POP off the diagram? Letters must touch each other horizontally, vertically, or diagonally.

Any P can be both the first and the last letter of the same POP.

Game 13

Timothy owns 13 chains. Each chain has a central link interlocked with two other links. There is a total of $13 \times 3 = 39$ links, all closed.

Timothy wants to use all the small chains to make a closed chain of 39 links. He needs 4 minutes to cut open a link and 10 minutes to close it by soldering.

Timothy makes the big chain in 140 minutes. How?

Game 14

$$5\ 5\ 5\ 5 = \quad 3$$
$$5\ 5\ 5\ 5 = \quad 5$$
$$5\ 5\ 5\ 5 = \quad 6$$
$$5\ 5\ 5\ 5 = \quad 26$$
$$5\ 5\ 5\ 5 = \quad 30$$
$$5\ 5\ 5\ 5 = \quad 50$$
$$5\ 5\ 5\ 5 = \quad 55$$
$$5\ 5\ 5\ 5 = 120$$

Add arithmetical symbols between the 5's to make every equation true.

Game 15

Timothy leaves Paris, driving at a constant speed. After a while he passes a "milestone"—actually a kilometer stone, of course —displaying a two-digit number. An hour later he passes a milestone displaying the same two digits, but in reversed order. In another hour he passes a third milestone, with the same two digits (backward or forward) separated by a zero.

What is the speed of Timothy's car?

Game 16

$$6\ 6\ 6\ 6 = \quad 5$$
$$6\ 6\ 6\ 6 = \quad 6$$
$$6\ 6\ 6\ 6 = \quad 8$$
$$6\ 6\ 6\ 6 = \quad 24$$
$$6\ 6\ 6\ 6 = \quad 30$$
$$6\ 6\ 6\ 6 = \quad 48$$
$$6\ 6\ 6\ 6 = \quad 66$$
$$6\ 6\ 6\ 6 = 180$$

Add arithmetical symbols between the 6's to make every equation true.

Game 17

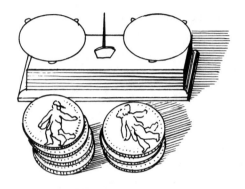

Of seven coins, which all look alike, five have the same weight while two are slightly heavier.

Using a balance of two pans, without weights, how many operations are necessary to tell which are the two heavy coins?

Game 18

$$7\ 7\ 7\ 7 = \quad 3$$
$$7\ 7\ 7\ 7 = \quad 8$$
$$7\ 7\ 7\ 7 = \quad 13$$
$$7\ 7\ 7\ 7 = \quad 15$$
$$7\ 7\ 7\ 7 = \quad 48$$
$$7\ 7\ 7\ 7 = \quad 49$$
$$7\ 7\ 7\ 7 = \quad 56$$
$$7\ 7\ 7\ 7 = 105$$

Add arithmetical symbols between the 7's to make every equation true.

Game 19

Timothy goes to a fountain with two containers, one of 8 liters, the other of 11 liters.

He wants to have twice as much water in one container as in the other. What is the shortest way to do it?

Game 20

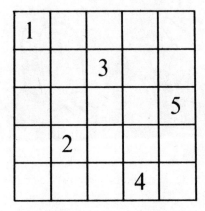

Complete this magic square with the integers from 6 through 25. Each horizontal, vertical, and diagonal must add up to 65.

Game 21

Timothy is in a bicycle race on a closed circuit.

After several hours of pedaling he realizes that 1/5 of the racers in front of him plus 5/6 of the racers in back of him add up to the total number of racers.

How many cyclists are there in the race?

Game 22

L

L E L

L E V E L

L E L

L

How many ways can you read LEVEL off the diagram, using letters that touch each other?

Any L or any E can be used twice in the same LEVEL.

Game 23

Timothy is riding a bicycle on a road that can be thought of as having four parts of equal length.

On the first fourth, which is level, he pedals at 10 kilometers per hour.

On the second fourth, an upward slope, he goes 5 kilometers per hour.

On the third fourth, a downward slope, he goes 30 kilometers per hour.

On the fourth fourth, which is level again—but with the wind pushing him—he goes 15 kilometers per hour.

What is Timothy's average speed?

Game 24

$$8 \ 8 \ 8 \ 8 = 10$$
$$8 \ 8 \ 8 \ 8 = 15$$
$$8 \ 8 \ 8 \ 8 = 56$$
$$8 \ 8 \ 8 \ 8 = 65$$
$$8 \ 8 \ 8 \ 8 = 80$$
$$8 \ 8 \ 8 \ 8 = 120$$
$$8 \ 8 \ 8 \ 8 = 192$$
$$8 \ 8 \ 8 \ 8 = 520$$

Add arithmetical symbols between the 8's to make every equation true.

Game 25

Timothy rents a car to drive to a city 100 kilometers away. He stops halfway and picks up a friend, who rides the last 50 kilometers with him.

Returning in the evening with his friend, Timothy drops him where he picked him up, then drives on to his starting point, where he is charged $24 for car rental.

Timothy and his friend share the expenses equitably. How much should each one pay?

Game 26

$$9 \ 9 \ 9 \ 9 = 7$$
$$9 \ 9 \ 9 \ 9 = 9$$
$$9 \ 9 \ 9 \ 9 = 10$$
$$9 \ 9 \ 9 \ 9 = 19$$
$$9 \ 9 \ 9 \ 9 = 80$$
$$9 \ 9 \ 9 \ 9 = 81$$
$$9 \ 9 \ 9 \ 9 = 90$$
$$9 \ 9 \ 9 \ 9 = 720$$

Add arithmetical symbols between the 9's to make every equation true.

Game 27

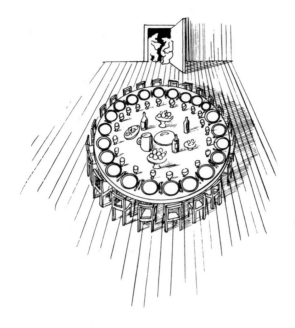

Every week 22 friends dine around a circular table. At each new meal, every diner tries to sit between two new friends. Thus if George sits between Peter and Paul this week he won't sit next to Peter or Paul next week or later, if he can avoid it.

After how many weeks will each diner have sat exactly once beside every other?

Game 28

$$1 \ 2 \ 3 \ 4 \ 5 \ 6 \ 7 \ 8 \ 9 = 100$$

Add arithmetical symbols on the left of the equals sign, between the numbers, to make this equation true. (There is an easy answer.)

NUMERICAL GAMES

Game 29

In a certain town, of each 100 men 85 are married, 70 have a telephone, 75 own a car, and 80 own their own house.

Always on a base of 100 men, what is the least possible number who are married, have a telephone, own a car, *and* own their own house?

Game 30

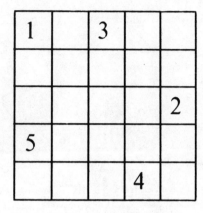

Complete this magic square with the integers from 6 through 25. Each horizontal, vertical, and diagonal, must add up to 65.

Game 31

A certain product is sold as either a liquid or a powder. A survey reveals that:

- 1/3 of the consumers interviewed do not use the powder;
- 2/7 do not use the liquid;
- 427 use both liquid and powder;
- 1/5 do not use the product at all.

How many consumers were interviewed?

Game 32

How many ways can you read SPICE off the diagram, using letters that touch each other?

Game 33

Timothy wants to divide in two equal parts the liquid filling a 16-liter container. To do it, he must use only the first container and two others, one of 11 liters and one of 6 liters.

How many operations are needed to achieve the division without spilling a drop of liquid?

Game 34

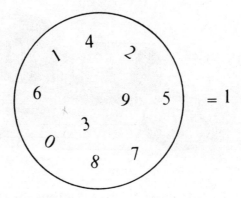

Use every digit in the circle once and only once, and insert simple arithmetical symbols to add up to 1.

Game 35

Two boats go up and down a river between two towns. They have the same two constant speeds: a high speed going downstream and a low speed going upstream.

The first boat leaves town A as the second boat leaves town B. They pass each other 7 miles from town A; they stop 4 minutes each at their destinations; they start back and pass each other the second time at 9 miles from town A.

What is the distance between the towns?

Game 36

$$\dots\dots\dots\dots\dots\dots\dots\dots\dots = 20$$
$$\dots\dots\dots\dots\dots\dots\dots\dots\dots = 20$$
$$\dots\dots\dots\dots\dots\dots\dots\dots\dots = 20$$
$$\dots\dots\dots\dots\dots\dots\dots\dots\dots = 20$$
$$\dots\dots\dots\dots\dots\dots\dots\dots\dots = 20$$
$$\dots\dots\dots\dots\dots\dots\dots\dots\dots = 20$$
$$\dots\dots\dots\dots\dots\dots\dots\dots\dots = 20$$
$$\dots\dots\dots\dots\dots\dots\dots\dots\dots = 20$$
$$\dots\dots\dots\dots\dots\dots\dots\dots\dots = 20$$

Arrive at 20 in nine different ways, using:
- on the first line, only 1's;
- on the second line, only 2's;
- and so on . . . ;
- on the ninth line, only 9's.

On every line, use the given digit no more than six times.

Game 37

A total of 15 delegates from Africa, Asia, America, and Europe meet at an international conference.

Each continent sends a different number of delegates, and each is represented by at least 1 delegate.

America and Asia send a total of 6 delegates.

Asia and Europe send a total of 7 delegates.

Which continent has sent 4 delegates?

Game 38

$$\star\star\star, 4\star\star \times 7 = 6{,}743{,}\star 56$$

Replace each star by a digit to make the equation true.

Game 39

Two 1-liter bottles are full of milk. Timothy owns two gauges of 40 and 70 centiliters. Using only these four containers, he wants to get 30 centiliters of milk in each gauge without spilling a drop of milk.

Timothy can do it in 6 operations. How?

Game 40

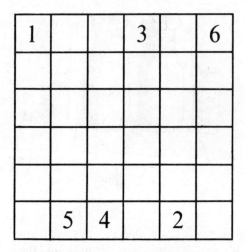

Complete this magic square with the integers from 7 through 36. Each horizontal, vertical, and diagonal must add up to 111.

Game 41

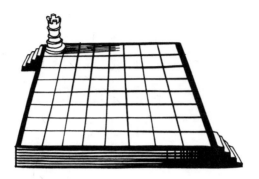

A rook crosses a chessboard from the northwest corner to the southeast corner. It can only move:

- south, east, or west;
- without entering or crossing any square more than once.

A rook can move any number of squares. How many different paths are possible?

Game 42

```
        1
      1 2 1
    1 2 1 2 1
      1 2 1
        1
```

How many ways can you read 12121 off the diagram, using digits that touch each other?

Any 1 or any 2 can be used several times in the same 12121.

Game 43

Timothy and Urban wish to compare their bicycling speeds, but they only have one bicycle. On a level road Timothy races from kilometer 1 to kilometer 12, Urban timing him from the back seat. On the same road Urban races from kilometer 12 to kilometer 24, with Timothy on the back seat.

Timothy wins easily. Is it because he rides faster, weighs more, or because of some other reason?

Game 44

$$X \ X \ X \ X$$
$$Y \ Y \ Y \ Y$$
$$Z \ Z \ Z \ Z$$
$$\overline{Y \ X \ X \ X \ Z}$$

Which three digits are represented by X, Y, and Z in this sum?

Game 45

Every day after work, Timothy takes the same train to his suburb. His wife meets him at the station and drives him home.

One day, without telling his wife, Timothy takes an earlier train and starts walking home. He passes his wife, who unfortunately does not see him. If she had, he would have been home 20 minutes earlier than usual. After he walks on for 25 minutes his wife, who has been to the station and back without losing any time, catches up with him. He gets in the car and arrives home at the usual time.

Timothy's wife drives at a constant speed; Timothy loses no time while stepping out of the train, getting in the car, and so on.

How much earlier was the train Timothy took on that day?

Game 46

```
              ★ ★ 8 ★ ★
          ┌─────────────────
    ★ ★  │ ★ ★ ★ ★ ★ ★ ★
          ★ ★ ★
          ───────
        0 0 0 ★ ★
              ★ ★
        ─────────
            0 ★ ★ ★
              ★ ★ ★
            ─────────
              0 0 1
```

In this division, which digits are hidden behind stars?

Game 47

There is 1 liter of water in container A and 1 liter of milk in container B.

Timothy pours 1/2 liter of water from container A into B, mixes carefully, and pours 1/4 liter of the mixture back into A.

Timothy now mixes carefully what is in A and pours 1/4 liter of the mixture into B. And finally, after a thorough mixing, he pours 1/2 liter of the mixture in B into A.

After these operations, is there more milk in A than there is water in B?

Game 48

Consider all the whole numbers from zero to 1 billion (1,000,000,000).

What is the sum of all the digits needed to write down these numbers?

Game 49

$$
\begin{array}{r}
X\ Y\ Z \\
+\ \ A\ B \\
\hline
C\ D\ E\ F
\end{array}
\qquad
\begin{array}{r}
X\ Y\ Z \\
-\ \ A\ B \\
\hline
B\ G\ A
\end{array}
$$

The same two numbers are added on the left and subtracted on the right.

Each letter represents a different digit. That is, if A is 3 (in all three positions), B cannot be 3.

Find all the digits.

Game 50

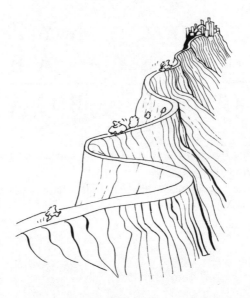

Three friends, Andrew, Bernard, and Claude, leave at the same time and place to go to town, 8 kilometers away.

Andrew starts on foot. Bernard takes Claude in his car. After a certain time Claude gets out of the car and goes on, walking. Bernard drives back to Andrew, takes him in the car, and drives him to town. All three arrive at the same time.

Andrew and Claude walk at a constant speed of 6 kilometers per hour. Bernard drives at a constant speed of 30 kilometers per hour.

How long did the trip last?

Game 51

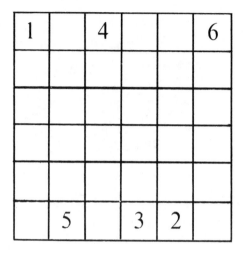

Complete this magic square with the integers from 7 through 36. Each horizontal, vertical, and diagonal must add up to 111.

(Hint: If the problem looks too tough, you can use the diagonal numbers in the answer to Game 40—they are the same in this game.)

Game 52

Timothy is preparing for a 42,000-kilometer trip in his car, a traditional four-wheel model.

Buying tires which each last 24,000 kilometers, Timothy contends that 7 will be enough.

Is he right?

Game 53

A wine merchant owns three casks of 100 liters, one of 50, and one of 25. The 100-liter casks are filled with Burgundy, Bordeaux, and Algerian wine respectively. The small casks are empty.

To make three qualities of wine, the merchant mixes the three wines, using the five casks exclusively.

After six operations, the three large casks are again full. The first is 1/4 Burgundy, 1/2 Bordeaux, and 1/4 Algerian. The second is 1/4 Bordeaux, 1/4 Algerian, and 1/2 Burgundy. The third is 1/2 Algerian, 1/4 Burgundy, and 1/4 Bordeaux.

What did the merchant do?

Game 54

```
            O
         O  L  O
      O  L  S  L  O
   O  L  S  O  S  L  O
      O  L  S  L  O
         O  L  O
            O
```

How many ways can you read OSLO off the diagram, using letters that touch each other?

Any O can be the first and the last letter of the same OSLO.

Game 55

Timothy is having dinner with a friend. He brought five dishes and his friend three dishes.

At the last minute another friend comes and eats with them.

The second friend pays $4 as his share. If all the dishes have the same value, how can the money be divided between Timothy and his first friend?

Game 56

```
      O E E
  ×     E E
  ─────────
    E O E E
    E O E
  ─────────
    O O E E
```

In this multiplication, each E stands for an even digit and each O stands for an odd digit.

There is only one possible solution. What is it?

Game 57

Timothy loves to walk. His constant speed is 6 kilometers per hour.

Every day at noon he meets a friend at an inn about halfway between their homes. The friend walks a little slower than Timothy: 5.5 kilometers per hour.

If both walkers reach the inn exactly at noon, how far are they from each other at 11:00?

Game 58

Timothy is at a water tap with containers of 19 liters, 13 liters, and 7 liters, all empty. Timothy wants to have exactly 10 liters in each of two containers, without spilling a drop.

How many operations are necessary?

Game 59

```
        P P P
  ×       P P
  _____
      P P P P
    P P P P
  _____
  P P P P P
```

In this multiplication, each P stands for a prime number less than 10—that is, 2, 3, 5, or 7.

What is the solution?

Game 60

In a stable there are men and horses. In all, there are 22 heads and 72 feet.

How many men and how many horses are in the stable?

Game 61

Nine schoolchildren form a circle. To choose a leader they decide to start from one of them, count up to 5 clockwise, ask the fifth player to leave, and so on. The last player left in the circle is the leader.

Andrew does the counting. He wants to take advantage of this to become the leader. Let's call him and his friends by the letters A through I, clockwise. With which child should he start to count?

Game 62

Timothy owns ten containers of 1, 2, 4, 5, 6, 12, 15, 22, 24, and 38 liters. He fills each one with one liquid only: one is full of milk, others are full of water, and still others are full of oil. But one container remains empty.

He has poured twice as much water as milk and twice as much oil as water.

What is in each container?

Game 63

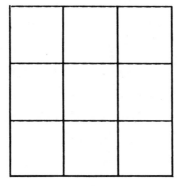

Is it possible to make a magic square with the first nine prime numbers? They are:

1 2 3 5 7 11 13 17 19

Game 64

Timothy sells cloth, fixing its retail price at 40% above its wholesale cost. He discovers that because the meterstick he has been using is not exact, he is only earning 39%.

How long is Timothy's "meter"? (Note: He uses it only to *sell* cloth; the amounts he buys are correct.)

Game 65

R
R E R
R E V E R
R E V I V E R
R E V E R
R E R
R

How many ways can you read REVIVER off the diagram, using letters that touch each other?

Any R, E, or V can be used twice in the same REVIVER.

Game 66

Timothy is on a ladder placed against a wall he is painting. He starts on the middle rung, goes up five rungs, down seven rungs, up four rungs, and up nine more rungs, to reach the top bar.

How many rungs are there on the ladder?

Game 67

The minute hand of a clock is out of its normal position, showing 5 more minutes than it should. Thus when the hour hand is exactly on 12, the minute hand is not on 12, but on 1.

It is near noon and the hour hand of the clock is exactly under the minute hand. What time is it?

Game 68

Timothy multiplied two five-digit numbers and wrote up the answer. Unfortunately, one digit (represented by a star in the drawing) was illegible.

$$98564 \times 54972 = 541\star260208$$

To determine the missing digit must Timothy do the whole multiplication again, or is there a shorter method?

Game 69

Looking for employment, Timothy hesitates between two jobs. The first offers $9,000 per year, with the promise of a $1,000 per year raise twice a year. The second also starts at $9,000 per year, with the promise of a $2,000 per year raise every year.

After reflection, Timothy chooses the first job. Why?

Game 70

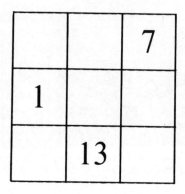

Is it possible to complete the magic square with other different prime numbers, each less than 100?

Game 71

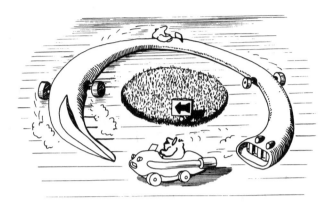

Timothy and Urban compete in a car rally, going several times around a closed circuit. Timothy can drive the circuit in 25 minutes, but Urban takes 30 minutes.

If the two drivers start at the same time, how long will it take Timothy to lap Urban?

Game 72

When Timothy is as old as his father is now, his sister will be twice as old as she is now and the age of his father will be twice the age Timothy will be when his sister is as old as his father is now.

The total age of the three is a century. How old is everyone?

Game 73

```
              L
            L A L
          L A V A L
        L A V A V A L
      L A V A L A V A L
        L A V A V A L
          L A V A L
            L A L
              L
```

How many ways can you read LAVAL off the diagram, using letters that touch each other? (Laval is an old town in the northwest of France.)

Any L or any A can be used twice in the same LAVAL.

Game 74

Timothy owns five bags of 20 coins each. Every coin should weigh 10 grams, but only the coins in three bags have the correct weight. Those of one bag weigh 9 grams each and those of another weigh 11 grams each.

In one weighing how can Timothy determine which bag contains the heavy coins and which bag contains the light coins? He has a weighing machine whose hand shows the exact weight on a dial.

Game 75

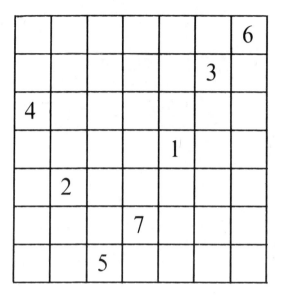

Complete this magic square with the integers from 8 through 49. Each horizontal, vertical, and diagonal must add up to 175.

Game 76

The numbers on a die are not placed at random. On every pair of opposite faces the sum of the numbers is 7: 6 is opposed to 1, 5 to 2, and 3 to 4.

There are several other ways of placing six numbers on a die. How many ways are there altogether?

Game 77

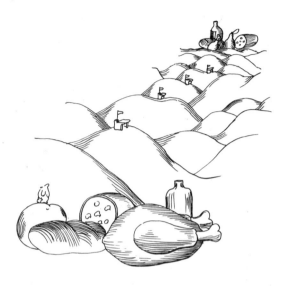

A traveler undertakes to walk, alone and without help, across a desert entirely lacking in resources. Every 20 kilometers on the 100-kilometer trail there is a shelter. (20 kilometers is exactly what a traveler can walk in 1 day.)

The traveler can only carry 3 days' food. He can stock food only in the shelters.

How many days will it take him to cross the desert?

Game 78

```
            A
          A N A
        A N N N A
      A N N A N N A
        A N N N A
          A N A
            A
```

How many ways can you read ANNA off the diagram, using letters that touch each other?

An A can both begin and end the same ANNA. But the two N's should be different.

Game 79

Timothy and Urban play a game with two dice. But they do not use the numbers. Some of the faces are painted red and the others blue.

Each player throws the dice in turn. Timothy wins when the two top faces are the same color. Urban wins when the colors are different. Their chances are even.

The first die has 5 red faces and 1 blue face. How many red and how many blue are there on the second die?

Game 80

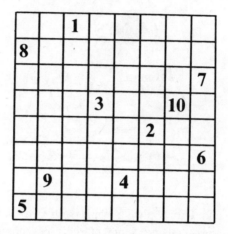

Complete this magic "chessboard" square with the integers from 11 through 64. Each horizontal, vertical, and diagonal must add up to 260.

NUMERICAL GAMES

Game 81

Timothy and two friends are digging identical holes in a field.

When Timothy works with Urban, they dig 1 hole in 4 days. When Timothy works with Vincent, they dig 1 hole in 3 days. When Urban works with Vincent, they dig 1 hole in 2 days.

When Timothy works alone, how long does it take him to dig 1 hole?

Game 82

An explorer undertakes to cross a desert with the help of porters. The length of the trail corresponds to 6 days' walk. But the explorer, like each porter, can carry only 4 days' food.

How many porters does the explorer need? In fact, can he cross at all?

Game 83

$$1$$
$$1 \; 2 \; 1$$
$$1 \; 2 \; 3 \; 2 \; 1$$
$$1 \; 2 \; 3 \; 3 \; 3 \; 2 \; 1$$
$$1 \; 2 \; 3 \; 3 \; 4 \; 3 \; 3 \; 2 \; 1$$
$$1 \; 2 \; 3 \; 3 \; 3 \; 2 \; 1$$
$$1 \; 2 \; 3 \; 2 \; 1$$
$$1 \; 2 \; 1$$
$$1$$

How many ways can you read 123343321 off the diagram, using digits that touch each other?

As usual, the same digit can be used twice in the same 123343321—but two 3's immediately following each other should be different.

Game 84

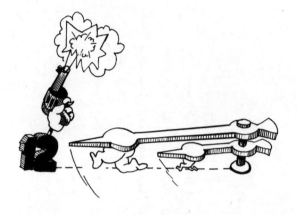

A clock runs fast. Its hands intersect every 61 minutes. How fast does the clock run in an hour?

Game 85

				42		
	35		14			
		28	21			
7						
	49					

Complete this magic square with the integers from 1 through 48 that are not multiples of 7. Each horizontal, vertical, and diagonal must add up to 175.

Game 86

Timothy collects whole numbers composed of four odd digits, all different. For example:

5713

Timothy knows how many such numbers there are. The first digit can be 9, 7, 5, 3, or 1. When it is chosen, the second digit can be any of the four remaining odd digits, and so on.

The total is:

$$5 \times 4 \times 3 \times 2 = 120$$

Timothy also knows the sum of the 120 odd numbers. What is it? (You don't actually have to add them all up.)

Game 87

A government issues two kinds of coins: one of 7 units and one of 11 units. Thus certain prices cannot be paid exactly, for example, 15 units.

What is the highest price that cannot be paid with any combination of the two coins?

Game 88

```
                1
              1 2 1
            1 2 1 2 1
          1 2 1 2 1 2 1
        1 2 1 2 1 2 1 2 1
          1 2 1 2 1 2 1
            1 2 1 2 1
              1 2 1
                1
```

How many ways can you read 12121 off the diagram, using digits that touch each other?

Any digit can be used several times in the same number.

Game 89

Timothy is an even-number fan. On a wall of his living room are four frames, each containing an even digit: 2, 4, 6, and 8. To reach absolute perfection, he wants to put the frames in such an order that the resulting four-digit number is a perfect square.

Is it possible?

Game 90

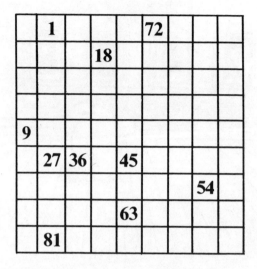

Complete this magic square with the integers from 2 through 80 that are not multiples of 9. Each horizontal, vertical, and diagonal must add up to 369.

Game 91

Timothy's balance is wrong: one of the arms is longer than the other. 1 kilogram on the left pan balances 8 melons on the right, while 1 kilogram on the right pan balances only 2 melons on the left.

If all the melons have the same weight, what is it?

Game 92

Timothy owns a good balance, but no weights. He decides to make his own weights by sawing a 121-gram bar of metal into a number of pieces. He has a system enabling him to weigh all whole numbers of grams from 1 through 121.

How does Timothy divide the bar? What is the least number of pieces necessary?

Game 93

D
D E D
D E I E D
D E I F I E D
D E I F I F I E D
D E I F I E I F I E D
D E I F I E D E I F I E D
D E I F I E I F I E D
D E I F I F I E D
D E I F I E D
D E I E D
D E D
D

How many ways can you read DEIFIED off the diagram, using letters that touch each other?

Any D, E, or I can be used twice in the same DEIFIED.

Game 94

The cashier of a bank pays out $1500, using a certain number of $1 bills, and ten times as many $5 bills, plus a certain number of $10 bills and twice as many $50 bills.

How many bills of each kind does he pay out?

Game 95

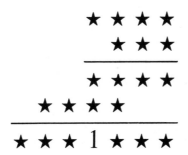

In this multiplication, each star represents either a 1 or a 0, for it is written in the binary system, where one is "1," two is "10," three is "1," and so on.

Only one digit is given. What is the value of each of the other stars?

(Note: In the binary system, $0 \times 0 = 0$, $0 \times 1 = 0$, and $1 \times 1 = 1$, just as in our decimal system; also $0 + 0 = 0$ and $0 + 1 = 1$, but $1 + 1 = 10$—that is, 0 and 1 "to carry.")

Game 96

In a small country each railroad station sells as many different kinds of tickets as there are other stations. The ticket for A to B is different from the ticket for B to A.

A new line goes to several new stations. The day it is opened 34 new tickets go on sale.

How many stations were there before? How many new stations have been opened?

Game 97

The French ambassador gives a reception. Half his guests are foreigners, whose official language is not French.

Each guest says "Bonjour" to the ambassador. And, to be polite, each guest says hello to every other guest in the official language of the person he is talking to.

The French ambassador answers "Soyez le bienvenu" to every guest.

In all, 78 "bonjours" are said. How many guests were there?

Game 98

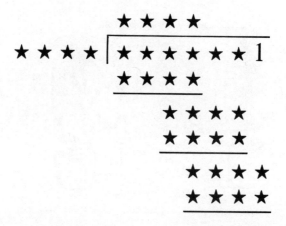

In this division, each star represents either 0 or 1, for it is written in the binary system.

Only one digit is given. What are the others?

Game 99

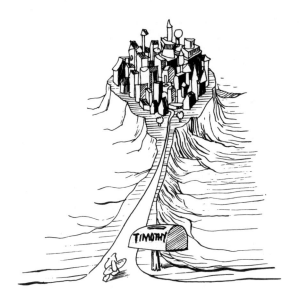

In Timothy's house there are several rooms. (A hall separated from the rest of the house by a door or doors is counted as a room.)

Each room has an even number of doors, including doors that lead outside.

Is the number of outside doors even or odd?

Game 100

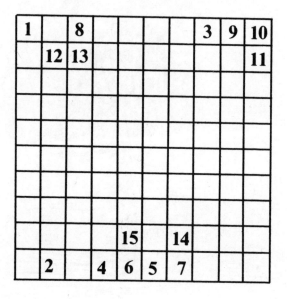

Complete this magic square with the integers from 16 through 100. Each horizontal, vertical, and diagonal must add up to 505.

SOLUTIONS

Game 1

Timothy entered each store with twice as much as the amount which is $1 more than he had when he left. Since he was broke when he left the fifth store, he entered it with:

$$(0 + 1) \times 2 = \$2$$

Likewise, he entered:

- the fourth store with $(2 + 1) \times 2 = \$6$;
- the third store with $(6 + 1) \times 2 = \$14$;
- the second store with $(14 + 1) \times 2 = \$30$;
- the first store with $(30 + 1) \times 2 = \$62$.

Timothy started with $62.

Game 2

At each vertex of the diamond, an A yields 1 ACE (that is, you can only get to 1 C from each A in a corner), or a total of 4.

On each side of the diamond, an A yields 2 ACE's (that is, you can get to 2 C's from each A on a side), or a total of 8.

Thus ACE can be read off in 12 ways.

Game 3

During the first 12 hours the minute hand is over the hour hand at 12:00 and won't cross it again for at least an hour. In the last hour the minute hand does not cross the small one except at the twelfth hour (12:00).

However, the hands cross once during each hour in between; the 12 crossings in 12 hours determine 11 intervals, and during each interval the hands form a right angle twice. Total: 22 right angles.

There is no overlap between the 12-hour periods, as far as right angles go. There are another 22 right angles in the second 12 hours. Thus the hands form 44 right angles during 24 hours.

Game 4

$$2 + 2 - 2 - 2 = 0$$
$$(2 \div 2) \times (2 \div 2) = 1$$
$$(2 \div 2) + (2 \div 2) = 2$$
$$(2 + 2 + 2) \div 2 = 3$$
$$2 + 2 + 2 - 2 = 4$$
$$2 + 2 + (2 \div 2) = 5$$
$$(2 \times 2 \times 2) - 2 = 6$$
$$(2 \times 2 \times 2) + 2 = 10$$
$$(2 + 2 + 2) \times 2 = 12$$

Game 5

The first train met is pulling in as our train leaves the station. The last train met is pulling out as our train arrives at its destination.

In between, we meet a train every half-hour—that is, 9 trains. Total trains met: 11.

Game 6

$$(3 + 3 + 3) \div 3 = 3$$
$$[(3 \times 3) + 3] \div 3 = 4$$

$$3 + 3 - (3 \div 3) = 5$$
$$3 + 3 + 3 - 3 = 6$$
$$3 + 3 + (3 \div 3) = 7$$
$$(3 \times 3) - (3 \div 3) = 8$$
$$(3 \times 3) + 3 - 3 = 9$$
$$(3 \times 3) + (3 \div 3) = 10$$

Game 7

At 5:00 the clock takes 6 seconds for 4 intervals between strikes: each interval is 1 1/2 seconds. The time needed to strike 9:00 confirms this calculation.

At any hour there are 1 less intervals than strikes. Intervals in 24 hours are:

$$2(0 + 1 + 2 + 3 + 4 + 5 + 6 + 7 + 8 + 9 + 10 + 11) = 132$$

Time needed for striking is:
$$1.5 \times 132 = 198 \text{ seconds}$$

Game 8

$$[(4 \times 4) - 4] \div 4 = 3$$
$$[(4 + 4) \div 4] + 4 = 6$$
$$4 + 4 - (4 \div 4) = 7$$
$$(4 \times 4) - 4 - 4 = 8$$
$$(4 \times 4) + 4 + 4 = 24$$
$$[(4 + 4) \times 4] - 4 = 28$$
$$(4 \times 4) + (4 \times 4) = 32$$
$$(4 + 4 + 4) \times 4 = 48$$

Game 9

Ten operations are needed. The table shows how many liters are in each container after each operation. In the first operation the first container is filled; in the second and third, the second container receives the water from the first, which is then refilled; in the fourth and fifth, the second container receives as much as possible from the first and is then emptied; and so on.

7-liter container	11-liter container
7	0
0	7
7	7
3	11
3	0
0	3
7	3
0	10
7	10
6	11

Game 10

1	12	7	14
8	13	2	11
10	3	16	5
15	6	9	4

Game 11

73 hens lay a total of 1 dozen eggs in 1 day, and 37 hens eat a total of 1 kilogram of wheat in 1 day. To get a dozen eggs you must feed 73 hens for 1 day. This requires 73/37 kilograms of wheat—a little less than 2.

Game 12

Let us count the POP's according to the position of the first P:
- at the center: P touches 4 O's each touching 4 P's, yielding a total of 16 words;
- in a corner: each of 4 P's touches 1 O and therefore 4 OP's, yielding a total of 16 words;
- on a side: each of 4 P's touches 2 O's, each touching 4 P's, yielding a total of 32 words.

In all there are 64 POP's.

Game 13

Timothy leaves 10 small chains intact and opens the 9 links of the last 3. With these 9 links he connects the 10 first chains. To close the long chain he need only open and close a last link.

Thus Timothy opens and closes 10 links, which is done in:

$$10 \times (10 + 4) = 140 \text{ minutes}$$

Game 14

$$(5 + 5 + 5) \div 5 = \quad 3$$
$$5 + [(5 - 5) \times 5)] = \quad 5$$
$$[(5 \times 5) + 5] \div 5 = \quad 6$$
$$(5 \times 5) + (5 \div 5) = \quad 26$$
$$[5 + (5 \div 5)] \times 5 = \quad 30$$
$$(5 \times 5) + (5 \times 5) = \quad 50$$
$$[(5 + 5) \times 5] + 5 = \quad 55$$
$$(5 \times 5 \times 5) - 5 = 120$$

Game 15

The number of kilometers on the first "milestone" can be written: $10A + B$

The number of kilometers on the second milestone is:
$$10B + A$$

On the third milestone, the number can be either $100A + B$ or $100B + A$.

Since Timothy's speed is constant the distances between the first and second milestones and between the second and third milestones are equal. Then the number of hundreds is 1 in the maximum case: $91 + (91 - 19) = 163$.

Is it $100A$ or $100B$ that equals 100? It has to be $100A$: the number on the first milestone, where A stands for tens, must be smaller than the number on the second milestone, where B stands for tens. Then we have:

$$(10B + 1) - (10 + B) = (100 + B) - (10B + 1)$$
$$B = 6$$

The milestones carry the numbers: 16   61   106

Timothy's speed is 45 kilometers per hour.

Game 16

$$[(6 \times 6) - 6] \div 6 = \quad 5$$
$$6 + [(6 - 6) \times 6] = \quad 6$$
$$6 + [(6 + 6) \div 6] = \quad 8$$
$$(6 \times 6) - 6 - 6 = \quad 24$$
$$[6 - (6 \div 6)] \times 6 = \quad 30$$
$$(6 \times 6) + 6 + 6 = \quad 48$$
$$[(6 + 6) \times 6] - 6 = \quad 66$$
$$[(6 \times 6) - 6] \times 6 = 180$$

Game 17

Let us designate the coins A through G.

Of coins A through F, at least one is heavy—perhaps two. We weigh ABC against DEF. If the pans balance, each group of three contains one heavy coin. In this case we weigh A against B (if they balance, C is heavy; if they don't, the pan that goes down holds the heavy coin), and then D against E (the same procedure). Total: three operations.

If the pans do not balance when weighing ABC against DEF —let us say ABC is heavier—it is because ABC contains one or two heavy coins. We weigh A against B. If A is heavier it is the first heavy coin, and the second will be found by weighing C against G. If the pans balance, the heavy coins are either A and B or C and G. The two pairs must be weighed against each other.

Again, only three operations are necessary.

Game 18

$$(7 + 7 + 7) \div 7 = \quad 3$$
$$[(7 \times 7) + 7] \div 7 = \quad 8$$
$$7 + 7 - (7 \div 7) = \quad 13$$
$$(7 \div 7) + 7 + 7 = \quad 15$$
$$(7 \times 7) - (7 \div 7) = \quad 48$$
$$7 \times 7 + 7 - 7 = \quad 49$$
$$[7 + (7 \div 7)] \times 7 = \quad 56$$
$$[(7 + 7) \times 7] + 7 = 105$$

Game 19

Here is how to have 4 and 8 liters in the containers, in 16 operations:

11-liter container	8-liter container
11	0
3	8
3	0
0	3
11	3
6	8
6	0
0	6
11	6
9	8
9	0
1	8
1	0
0	1
11	1
4	8

Game 20

1	14	22	10	18
7	20	3	11	24
13	21	9	17	5
19	2	15	23	6
25	8	16	4	12

Game 21

Since the track is a closed circuit, the racers in front of Timothy are also in back of Timothy; so either fraction can be multiplied by all the racers except Timothy. In fact:

$$\frac{1}{5} + \frac{5}{6} = \frac{31}{30}$$

Timothy sees 30 competitors, so there are 31 cyclists in all.

Game 22

We first count the VEL's.

V touches 4 E's, each touching 3 L's, which yields 12 VEL's.

Since LEV can be read 12 times too, there are 144 ways to read LEV VEL, and for each of these there is 1 LEVEL. So the answer is 144.

Game 23

Let L be the length in kilometers of each fourth of the road. Timothy rides:

- the first fourth in $\frac{L}{10}$ hours;

- the second fourth in $\frac{L}{5}$ hours;

- the third fourth in $\frac{L}{30}$ hours;

- the fourth fourth in $\frac{L}{15}$ hours.

The total time is:

$$\frac{L}{10} + \frac{L}{5} + \frac{L}{30} + \frac{L}{15} = \frac{2L}{5}$$

Since speed equals distance divided by time, we divide the length of the whole road, 4L, by $\frac{2L}{5}$ to get Timothy's average speed: 10 kilometers per hour.

Game 24

$$[(8 + 8) \div 8] + 8 = 10$$
$$8 + 8 - (8 \div 8) = 15$$
$$[8 - (8 \div 8)] \times 8 = 56$$
$$(8 \times 8) + (8 \div 8) = 65$$
$$(8 \times 8) + 8 + 8 = 80$$
$$[(8 + 8) \times 8] - 8 = 120$$
$$(8 + 8 + 8) \times 8 = 192$$
$$(8 \times 8 \times 8) + 8 = 520$$

Game 25

Timothy rode 200 kilometers and his friend 100 kilometers, for a total of 300 "passenger-kilometers." Each passenger-kilometer is worth:

$$\frac{2400}{300} = 8 \text{ cents}$$

Timothy should pay $16 and his friend $8.

Why bother about passenger-kilometers for such a simple problem? Because more often than not a wrong solution is considered: Timothy travels alone half of the trip, for which he should pay $12. He shares his car with a friend for the other half, for which each should pay $6, so Timothy should pay $18 and his friend only $6. But this assumes the use of the car has the same total value per kilometer to the riders with two passengers as with one.

Game 26

$$9 - [(9 + 9) \div 9] = 7$$
$$9 - [(9 - 9) \times 9] = 9$$
$$[(9 \times 9) + 9] \div 9 = 10$$
$$(9 \div 9) + 9 + 9 = 19$$
$$(9 \times 9) - (9 \div 9) = 80$$
$$[(9 \times 9) + 9] - 9 = 81$$
$$[9 + (9 \div 9)] \times 9 = 90$$
$$(9 \times 9 \times 9) - 9 = 720$$

Game 27

The problem is a trap. Each diner has two new friends beside him every week. But he has 21 friends, so he can't sit beside all of them in a whole number of weeks.

Game 28

$$1 + 2 + 3 + 4 + 5 + 6 + 7 + (8 \times 9) = 100$$

Game 29

On the base of 100 men:

- 15 are not married;
- 30 do not have a telephone;
- 25 do not have a car;
- 20 do not own their own house.

It is possible that these 90 men are all different, which would leave only 10 men with wife, phone, car, and house.

Game 30

1	13	3	25	23
21	18	11	6	9
22	19	12	10	2
5	8	15	20	17
16	7	24	4	14

Game 31

Divide the respondents into four sets:

A: do not use the product
B: use liquid only
C: use both liquid and powder
D: use powder only

We know that:

$A + B$ is 1/3 of the total
$A + D$ is 2/7 of the total
A is 1/5 of the total
C is 427

Then:

B is $1/3 - 1/5 = 2/15$ of the total
D is $2/7 - 1/5 = 3/35$ of the total
$A + B + D$ is 44/105 of the total
C is the remaining 61/105 of the total

The number of consumers interviewed is:

$$427 \div \frac{61}{105} = 735$$

Game 32

S touches 4 P's. Each P touches:

- 1 I that touches 7 CE's;
- 2 I's, each touching 4 CE's.

There is a total of:

$$4(7 + 8) = 60 \text{ words}$$

Game 33

An even split can be achieved in 14 operations:

16-liter container	11-liter container	6-liter container
(16)	(0)	(0)
10	0	6
10	6	0
4	6	6
4	11	1
15	0	1
15	1	0
9	1	6
9	7	0
3	7	6
3	11	2
14	0	2
14	2	0
8	2	6
8	8	0

Game 34

$$\frac{35}{70} + \frac{148}{296} = 1$$

Game 35

Since the boats arrive and depart from their home towns at the same times, the problem is symmetrical. A distance from town A can be considered as if it were a distance from town B.

Thus we can add 7 miles from town A to 9 miles from town B to get the total distance, 16 miles.

Game 36

$$11 + 11 - 1 - 1 = 20$$
$$22 - 2 = 20$$
$$3^3 - (3 + 3 + \frac{3}{3}) = 20$$
$$(4 \times 4) + 4 = 20$$
$$(5 \times 5) - 5 = 20$$
$$6 + 6 + 6 + \frac{6 + 6}{6} = 20$$
$$7 + 7 + 7 - \frac{7}{7} = 20$$
$$8 + 8 + \frac{8 \times 8}{8 + 8} = 20$$
$$9 + 9 + 9\frac{+9}{9} = 20$$

Game 37

How many delegates are from Asia? Since they make a total of 6 with those from America which has sent at least 1, they can be 1, 2, 3, 4, or 5.

3 is impossible, since it would yield an equal number of delegates from America.

1 would yield 5 from America, 6 from Europe, and hence 3 from Africa. This is impossible, for one continent has sent 4 delegates.

2 would yield 4 from America, 5 from Europe, and 4 from Africa, which is impossible, for America and Africa would have sent the same number.

5 would yield 1 from America, 2 from Europe, and 7 from Africa. Still impossible: no number 4.

The possibility, 4 from Asia, 2 from America, 3 from Europe, and 6 from Africa, gives the answer: Asia.

Game 38

The ones digit of the first number multiplied by 7 must give a number ending in 6. Only 8 will do it. So we have:

$$***, 4*8 \times 7 = 6{,}743{,}*56$$

8 times 7 is 56. Since the product already ends in 56, the tens digit of the first number, times 7, must end in 0. Therefore, the tens digit is 0. We have:

$$***, 408 \times 7 = 6{,}743{,}*56$$

Now 408 times 7 is 2856. The missing digit in the product is 8:

$$***, 408 \times 7 = 6{,}743{,}856$$

The thousands digit times 7 must end in 1, since 1 plus "2 to carry" is the 3 in the product. The only multiple of 7 ending in 1 is 21, so the thousands digit is 3:

$$**3, 408 \times 7 = 6{,}743{,}856$$

We repeat this procedure to obtain the remaining digits:

$$963{,}408 \times 7 = 6{,}743{,}856$$

Game 39

Here are the six operations:

First bottle	Second bottle	40-centiliter gauge	70-centiliter guage
(100)	(100)	(0)	(0)
100	60	40	0
100	0	40	60
90	0	40	70
90	40	0	70
90	40	40	30
100	40	30	30

Game 40

1	32	34	3	35	6
30	8	28	27	11	7
19	17	15	16	20	24
18	23	21	22	14	13
12	26	9	10	29	25
31	5	4	33	2	36

Game 41

The rook can cross the uppermost east-west line on any of its 8 segments; after moving west or east (or not moving west or east, in effect, by moving more than 1 square south at a time) it can cross the next east-west line on any of its 8 segments; and so on. There are 7 east-west lines, and the crossings are independent of each other, so there are 8^7, or:

$$8 \times 8 \times 8 \times 8 \times 8 \times 8 \times 8 = 2,097,152 \text{ paths}$$

Game 42

We will consider the central "1" in 12121:

- In the center of the diagram, 1 touches 4 "2's", each touching 4 "1's". This makes 16 "121's" and 256 "12121's."
- Each corner "1" touches one "2" touching 4 "1's", which makes it the center of 16 "12121's". Total: 64 "12121's."
- Each "1" on a side touches 2 "2's," each touching 4 "1's," which makes it the center of 64 "12121's." For the 4 sides: 256 "12121's."

The grand total is 576 numbers.

Game 43

It is 11 kilometers from kilometer 1 to kilometer 12, but 12 kilometers from kilometer 12 to kilometer 24. Timothy bicycled a shorter distance.

Game 44

Three digits added together give a maximum of 27. Then Y, the first digit of the total, is 1 or 2.

Since the first digits add up to Z, the sum of X and Y ends in 0. Then $X + Y = 10$. Since Y is 1 or 2, there are two possibilities: $Y = 1, X = 9$

$$Y = 2, X = 8$$

There is 1 to carry to the tens column. Canceling the X's, $1 + Y + Z = 10$. Then $Y + Z$ ends in 9—and must be 9, since no two digits add up to 19. Now we fill out the two solutions:

$$Y = 1, X = 9, Z = 8$$
$$Y = 2, X = 8, Z = 7$$

They yield:

9 9 9 9	8 8 8 8
1 1 1 1	2 2 2 2
8 8 8 8	7 7 7 7
1 9 9 9 8	2 8 8 8 7

Only the first addition is correct.

Game 45

Timothy could have saved 20 minutes on his usual trip if his wife hadn't passed him; this 20 minutes is the time she needed to go to the station and back from the passing point, which is 10 minutes from the station by car.

It took her 25 minutes to drive from the passing point to the station and back to where she picked Timothy up. The pickup point is 5 minutes by car beyond the passing point. Since Timothy walked this distance in 25 minutes, he goes five times as slowly as the car; therefore, between the station and the passing point he walked for 50 minutes. He was at the passing point 50 minutes after the arrival of the early train and 10 minutes before the arrival of the usual train. There is 1 hour between the two trains.

Game 46

After the second and fourth steps it has been necessary to bring down a second digit from the dividend. Hence the quotient has two 0's: *080*.

The divisor times 8 is a two-digit number. The divisor can't be more than 12, for $8 \times 13 = 104$. But if it was less than 12, it couldn't give a three-digit number (first step) when multiplied by a digit. So the divisor is 12 exactly.

The divisor only gives a three-digit number when multiplied by 9. Then the quotient begins with 9 and the dividend begins with $108 = 9 \times 12$.

Similarly, the last digit of the quotient is 9; the last product, like the first, is 108 and is subtracted from 109 to give 1. The second product is, of course, 96.

The division becomes:

```
            90809
     12)108**09
        108
        ────
        000**
          96
         ────
         0109
          108
         ────
          001
```

The dividend, then, must be 1,089,709.

Game 47

No calculations are needed for this problem. Container A holds the same amount of liquid that it held before the operations; the same is true of B. Whatever milk A has acquired it has given up to B in water. Hence there is exactly as much milk in the water as there is water in the milk.

Game 48

Let us group the numbers in pairs:

0 and 999,999,999

1 and 999,999,998

2 and 999,999,997

and so on.

For each pair the digits add up to 81, and there are 500,000,000 pairs. The total of their digits is:

$$81 \times 500,000,000 = 40,500,000,000$$

But we have left out the last number, 1,000,000,000, whose digits add up to 1. So the grand total is: 40,500,000,001.

Game 49

We begin with the addition. The total has four digits; it can only begin with 1. In fact, it can only begin with 10, and X is 9:

```
  9 Y Z        9 Y Z
+   A B      −   A B
─────────    ─────────
1 0 E F        B G A
```

In the answer to the subtraction, B is 9 or 8. By the rules of the game it cannot be 9, since X is 9, so it is 8.

We note that A is greater than Y and A + Y is greater than 8. (Otherwise one of the results would begin with 9). In fact A + Y is greater than 10 (since E cannot be 0 or 1). Trial and error gives the only possible result:

```
  9 4 5        9 4 5
+   7 8      −   7 8
─────────    ─────────
1 0 2 3        8 6 7
```

Game 50

Andrew walked L kilometers on foot. Since the three travelers arrived at the same time, Andrew and Claude must have walked the same distance; since Andrew and Claude walk at the same speed, they walked an equal amount of time.

Claude walked L/6 hours. During this time, Bernard drove (8 — 2L) kilometers back to meet Andrew (the length of the road, minus L walked by Andrew and L walked by Claude), and (8 — L) kilometers with Andrew to town; in all, (16 — 3L) kilometers, which took him:

$$\frac{16 - 3L}{30} \text{ hours}$$

Since Bernard took the same time to reach town after dropping Claude that Claude took:

$$\frac{L}{6} = \frac{16 - 3L}{30}$$

and L = 2 kilometers.

The total time of the trip was:

$$\frac{L}{6} + \frac{8 - L}{30}$$

which is the time Andrew walked plus the time it took Bernard to drive him (8 — L) kilometers to town. Substituting 2 for L, the time was 32 minutes.

Game 51

1	32	4	33	35	6
12	8	28	27	11	25
19	23	15	16	14	24
18	17	21	22	20	13
30	26	9	10	29	7
31	5	34	3	2	36

Game 52

Yes. The trip is $42,000 \times 4 = 168,000$ "tire-kilometers" and the purchase is $24,000 \times 7 = 168,000$ tire-kilometers.

Here is how he rotates them: Let us number the tires from 1 to 7. Timothy starts using tires 1, 2, 3, and 4 on the wheels. He stops every 6,000 kilometers to put successively on the wheels:

$$2\ 3\ 4\ 5$$
$$3\ 4\ 5\ 6$$
$$4\ 5\ 6\ 7$$
$$5\ 6\ 7\ 1$$
$$6\ 7\ 1\ 2$$
$$7\ 1\ 2\ 3$$

Each tire has been used for 24,000 kilometers.

Game 53

The merchant does not need the 25-liter cask. Let us call the three big casks A, B, and C, and the 50-liter cask X. The six operations are:

- fill X from A;
- fill A from B (A is 1/2 Burgundy, 1/2 Bordeaux);
- fill B from C (B is 1/2 Bordeaux, 1/2 Algerian);
- fill C from A (C is 1/4 Burgundy, 1/4 Bordeaux, 1/2 Algerian);
- fill A from B (A is 1/4 Burgundy, 1/2 Bordeaux, 1/4 Algerian);
- fill B from X (B is 1/2 Burgundy, 1/4 Bordeaux, 1/4 Algerian).

Game 54

We select the place of the first O:

- at the center: it touches 4 S's, each touching 3 L's, each touching 3 O's, which yields 36 words;
- at the corners: no words;
- on the sides: each of the 8 O's touches 1 S, touching 3 L's, each touching 3 O's, which yields 72 words.

In all, there are 108 words.

Game 55

Since the second friend paid $4, the total cost of the meal must be $4 \times 3 = \$12$.

Eight dishes have been eaten. Each one costs $1.50.

Timothy brought $5 \times 1.5 = \$7.50$ worth of food. His share being $4, he receives $3.50.

The first friend brought $3 \times 1.5 = \$4.50$ worth of food and receives 50 cents.

Game 56

$$
\begin{array}{r}
3\,4\,8 \\
2\,8 \\
\hline
2\,7\,8\,4 \\
6\,9\,6 \\
\hline
9\,7\,4\,4 \\
\end{array}
$$

(The first step in the solution was that the multiplicand must begin with odd digit 1 or 3 to give a three-digit product when multiplied by an even digit. But the multiplicand gives a four-digit product beginning with at least 2 when multiplied by another even digit, so the multiplicand cannot begin with 1.)

Game 57

They are $6 + 5.5 = 11.5$ kilometers from each other.

Game 58

It can be done in 18 operations:

19-liter container	13-liter container	7-liter container
0	0	7
0	13	7
7	13	0
19	1	0
12	1	7
12	8	0
5	8	7
5	13	2
18	0	2
18	2	0
11	2	7
11	9	0
4	9	7
4	13	3
17	0	3
17	3	0
10	3	7
10	10	0

Game 59

```
    7 7 5
      3 3
  2 3 2 5
2 3 2 5
2 5 5 7 5
```

(Did you get the first step of the solution? One of the ones digits of the multiplicand and multiplier must be a 5, because $2 \times 2 = 4$, $2 \times 3 = 6$, $2 \times 7 = 14$, $3 \times 3 = 9$, $3 \times 7 = 21$, and $7 \times 7 = 49$, and none of these products ends in 2, 3, 5, or 7.)

Game 60

If there were only men in the stable, there would be $22 \times 2 = 44$ feet. But there are 72 feet. So the remaining 28 feet, or 14 pairs, account for 14 horses. Since each horse has 2 pairs of feet, 14 horses have $28 \times 2 = 56$ feet, leaving 16 feet belonging to men. Thus there are 8 men and 14 horses in the stable.

Game 61

Let us first suppose that Andrew starts counting with himself. The numbers under the letters show in which order the schoolchildren leave the circle.

A	B	C	D	E	F	G	H	I
2	8	5	4	1	6	3	9	7

If Andrew starts with A, the leader will be H. To be the leader himself he must start two children more clockwise, that is, with C.

Game 62

The volumes of milk, water, and oil are in the ratio 1:2:4. The sum of the three numbers being 7, the total volume of liquid must be a multiple of 7 liters.

The volumes of the ten containers add up to 129, which is a multiple of 7 plus 3. The empty container must have a capacity of a multiple of 7 plus 3. It will be 24 liters or 38 liters.

If the 38-liter container remains empty, the total volume of liquid will be 91 liters, including 13 liters of milk, which is not feasible.

Then, the 24-liter container remains empty, which yields the only solution:
- milk: 15-liter container;
- water: 1-, 2-, 4-, 5-, 6- and 12-liter containers;
- oil: 22- and 38-liter containers.

Game 63

It is impossible to make a magic square with the first nine prime numbers.

Their sum is 78. Since there are three horizontal lines, the sum of each one must be a third of 78, or 26. But eight of the nine primes are odd numbers; two lines must add up to 26 using three odd numbers, which is impossible.

Game 64

Say Timothy buys his cloth at $1 per meter, and let L be the actual length of his meterstick.

When Timothy thinks he is selling 1 meter of cloth he is

really selling L meters, which cost him $L, for $1.40, earning a profit of $(1.40-L). We have:

$$\frac{39}{100} = \frac{1.40 - 1}{L}$$

$$L = \frac{140}{139} = 1.0072$$

His "meter" is too long by 7.2 millimeters.

Game 65

Let us count how many times IVER can be read from the I. The I touches 4 V's, each touching 7 ER's, which yields 28 IVER's.

REVI can be read the same number of times, 28, so REVI IVER (and therefore REVIVER) can be read the square of 28 times, that is, 784 times.

Game 66

Say Timothy starts on rung O. He goes up to rung 5, down to rung (−2), up to rung 2, and again up to rung 11.

There are $2 \times 11 + 1 = 23$ rungs.

Game 67

At 11:55 the hour hand is between:
- the point where the minute hand should be, 55;
- the point where the minute hand really is, 60.

Then an intersection took place between 11:50 and 11:55. This is the intersection we want, since the next one occurs more than an hour later.

The real number of minutes after 11:00 is:

$$50 + N$$

where N is less than 5.

The minute hand shows—wrongly—(55 + N) minutes. The hour hand is on (55 + N) too.

The real time, calculated from the hour hand, is 11:00 plus 12N minutes, since N minutes have passed since the hour hand was at 55. (Caution: The hour hand of a clock moves 12 times as slowly as the minute hand, not 60 times as slowly.) Since the real time also equals 11:50 + N, we have:

$$12N = 50 + N$$

$$N = \frac{50}{11}$$

N is 4 minutes 33 seconds. The time is 11:54:33.

Game 68

Timothy only needs to check the multiplication, using the rule of 9.

The sum of the digits of the first number is 32, and the sum of 3 and 2 is 5. The sum of the digits of the second number is 27, and 2 + 7 = 9.

$$5 \times 9 = 45$$

The sum of the digits of 45 is 9. The readable digits add up to 28, then 10, then 1. The illegible digit must be 8, since 1 + 8 = 9.

(The rule of 9 is based on the fact that any power of 10 gives a remainder of 1 when divided by 9. This is why 40, 400, 4000, 40,000, and so on, all give a remainder of 4 when divided by 9. Adding the digits "squeezes out" the powers of 10, leaving remainders of 9 which obey the same rules of multiplication that the original large numbers did.)

Game 69

Let us examine the cumulative earnings offered by each job at 6-month intervals.

First job:

6 months:	$4,500 = $4,500	
12 months:	$4,500 + 5,000 =	$9,500
18 months:	$9,500 + 5,500 = $15,000	
24 months:	$15,000 + 6,000 = $21,000	
30 months:	$21,000 + 6,500 = $27,500	

Second job:

6 months:	$4,500 = $4,500	
12 months:	$4,500 + 4,500 =	$9,000
18 months:	$9,000 + 5,500 = $14,500	
24 months:	$14,500 + 5,500 = $20,000	
30 months:	$20,000 + 6,500 = $26,500	

The earnings offered by the first company are consistently higher than those offered by the second.

Game 70

43	61	7
1	37	73
67	13	31

Game 71

25 minutes after starting Timothy finishes the first lap and Urban has only driven 25/30 of the circuit, or 5/6. Hence Timothy gains 1/6 lap on Urban every 25 minutes. He laps Urban in 6 × 25 minutes = 150 minutes—that is, 2 1/2 hours.

Game 72

Let us call the ages of Timothy, his sister, and his father T, S, and F. We know that:

$$T + S + F = 100$$

The first part of the first sentence of the problem yields:

$$S + (F - T) = 2S$$

or

$$F - S - T = 0$$

Combining this with the first equation:

$$2F = 100$$

$$F = 50$$

The second part of the first sentence of the problem yields:

$$F + (F - T) = 2 [T + (F - S)]$$

$$3T - 2S = 0$$

But $T + S = 50$. Then Timothy is 20, his sister is 30, and his father is 50.

Game 73

We will count the words by the placement of V. All the V's form a diamond:

- in a corner of the diamond: each of 4 V's touches 8 AL's; the 8 VAL's yield $4 \times 8 \times 8 = 256$ words.
- on a side of the diamond: each of 4 V's touches 6 AL's; the 6 VAL's and 6 LAV's yield 144 words.

In all, LAVAL can be read 400 times.

Game 74

Timothy puts 31 coins on the scale. If we call the bags A, B, C, D, and E, he chooses these coins:

> 1 from A
> 2 from B
> 4 from C
> 8 from D
> 16 from E

Note that all the differences between the number of coins from two bags are different: 1, 2, 3, 4, 6, 7, 8, 12, 14, and 15. Adding plus and minus signs we have 20 differences corresponding to the 20 possible conditions of the bags.

For example:

- if A is light and B is heavy the weight is:

$$310 - 1 + 2 = 311$$

- if A is heavy and B light:

$$310 + 1 - 2 = 309$$

- if A is light and C heavy:

$$310 - 1 + 4 = 313$$

And so on.

Game 75

42	18	29	9	45	26	6
20	35	11	43	23	3	40
4	36	16	31	12	48	28
33	13	49	25	1	37	17
22	2	38	19	34	14	46
10	47	27	7	39	15	30
44	24	5	41	21	32	8

Game 76

We can place the number 1 anywhere. Then there are five ways of choosing the number which will be opposite the 1.

Imagine that the 1 and its opposite are on the top and bottom of the die. We still have to place the four remaining numbers on the four remaining faces around the die. Let X be one of these numbers. There are three ways of choosing its opposite number. Then there are two ways of placing the last two numbers.

The total number of ways is:

$$5 \times 3 \times 2 = 30$$

None of these 30 dice can be turned into one of the others by rotating it in any direction.

Game 77

Let the four shelters be A, B, C, and D.

During days 1 through 8 the traveler makes four round trips to A, each time leaving 1 day's food in the shelter. On day 9 he arrives with 2 more days' food, which makes 6.

During days 10 and 11 he makes one round trip to B, leaving 1 day's food. On day 12 he arrives with 2 more days' food, which makes 3. With this, he can reach his goal in 3 days. He crosses the desert in 15 days.

Game 78

We will count the words by the placement of the first A:

- in the center: it touches 4 N's horizontally or vertically, each making 15 ANNA's, yielding 60 words; it also touches 4 N's diagonally, each making 12 ANNA's, yielding 48 words;
- in a corner each of 4 A's makes 9 ANNA's, yielding 36 words;
- on a side: each of 8 A's makes 30 ANNA's, yielding 240 words.

In all we have:

$$
\begin{array}{r}
60 \\
48 \\
36 \\
\underline{240} \\
384
\end{array}
$$

Game 79

Each die has 6 faces. When two dice are thrown, there are 36 equally possible results. For chances to be even, there must be 18 ways of getting the same color on top.

Let X be the number of red faces on the second die. We have:

$$18 = 5X + 1 (6 - X)$$

$$X = 3$$

The second die must have 3 red faces and 3 blue faces.

Game 80

30	12	1	50	23	40	45	59
8	18	27	44	13	62	55	33
34	56	61	14	43	28	17	7
57	47	52	3	38	21	10	32
60	46	39	24	49	2	11	29
35	53	42	25	64	15	20	6
31	9	22	37	4	51	48	58
5	19	16	63	26	41	54	36

Game 81

In 1 day:

- Timothy and Urban dig 1/4 hole;
- Timothy and Vincent dig 1/3 hole;
- Urban and Vincent dig 1/2 hole.

Let us use Timothy's twin brother who would do the same amount of work if he were here. If Timothy, his brother, Urban, and Vincent work together for 1 day, they dig:

$$1/4 + 1/3 = 7/12 \text{ hole.}$$

Since Urban and Vincent dig 6/12 per day, Timothy and his brother dig 1/12. Then Timothy alone digs 1/24 hole.

Timothy can dig a hole in 24 days.

Game 82

The explorer can cross the desert using only two porters.

They all leave one morning, each with 4 days' food. At the end of the first day, each has 3 days' food left. The first porter goes back with 1 day's food, leaving 4 days' food each for the explorer and the second porter.

At the end of the second day, the two men each have 3 days' food. The second porter goes back with 2 days' food. The explorer goes on with 4 days' food, enough to reach his goal.

Game 83

Let us count the numbers "43321" from the central "4."

- The "4" touches horizontally and vertically 4 "3's," each touching 5 other "3's." Of these last "3's," one touches 7 "21's," two touch 2 "21's," and two touch 4 "21's" which yields 19 possible ends and a total of 4 × 19 = 76 "43321's";
- The "4" touches diagonally 4 "5's," each touching 4 other "3's." Of these two touch 2 "21's," and two touch 7 "21's," which yields 18 possible ends and a total of 72 "43321's."

Thus we have 148 possible ways of reading "43321" and, therefore, 148 ways of reading "12334." The grand total is the square of 148, or 21,904.

Game 84

We must first calculate how long it is between intersections on a normal clock. Say the first intersection is at noon: the next will be N minutes after 1:00. During these N minutes the minute hand, starting from O, advances to N minutes, while the hour hand, starting from 5 minutes and moving 12 times as slowly, advances to $5 + \dfrac{N}{12}$ minutes:

$$5 + \frac{N}{12} = N$$

$$5 = \frac{11N}{12}$$

$$N = \frac{60}{11} \text{ minutes}$$

The total time between two intersections is:

$$60 + \frac{60}{11} = \frac{720}{11} \text{ minutes}$$

The fast clock only takes 61 minutes. In 60 minutes it shows $\dfrac{60}{61}$ of $\dfrac{720}{11}$, or approximately 64 minutes 23 seconds.

The clock runs 4 minutes 23 seconds fast every hour.

Game 85

46	1	2	3	42	41	40
45	35	13	14	32	31	5
44	34	28	21	26	16	6
7	17	23	25	27	33	43
11	20	24	29	22	30	39
12	19	37	36	18	15	38
10	49	48	47	8	9	4

Game 86

Imagine that the 120 numbers are written in one long column for addition. When you add the right-hand digits, how many 1's will there be? 24, because there are 24 ways to choose the three other digits when one is already chosen. So the sum of the right-hand column is:

$$24 (1 + 3 + 5 + 7 + 9) = 600$$

All four columns have the same sum.
The grand total is:

```
    6 0 0
    6 0 0
    6 0 0
    6 0 0
  ─────────
  6 6 6 6 0 0
```

Game 87

The first kind of coin alone can pay these prices:

11 22 33 44 55 . . .

With one coin of 7 units, it pays:

7 18 29 40 51 . . .

With two 7-unit coins:

14 25 36 47 58 . . .

And so on . . .

And with ten 7-unit coins:

70 81 92 103 114 . . .

These are 11 different series. (The next series, with 11 coins of 7 units, repeats the first series.) The 11 series all have different terms, for 7 and 11 are mutually prime. Thus the 11 series include every integer of 70 or more.

But the 11th series fills the gaps left by the ten first ones, and the first missing number is the one that would come before 70 in the 11th series:

$$70 - 11 = 59$$

The highest nonpayable price is 59 units.

Game 88

We will count the numbers "12121" by the placement of the middle "1" in "12121." It can be:

- at the center or on the central diamond, which has 3 "1's" on a side (9 places). Each "1" touches 4 "2's" which each touch 4 "1's." It begins 16 "121's," ends 16 "121's," and is the center of $16 \times 16 = 256$ "12121's." Then the 9 "1's" yield 2304 numbers
- at a corner of the outer diamond (4 places). Each "1" touches 1 "2" which touches 4 "1's." It begins 4 "121's," ends 4 "121's," and is the center of 16 "12121's." The 4 "1's" yield 64 "12121's"
- on a side of the outer diamond (12 places). Each "1" touches 2 "2's," each of which touches 4 "1's." It begins 8 "121's," ends 8 "121's," and is the center of 64 "12121's." The 12 "1's" yield 768 "12121's."

The grand total is 3136 ways of reading "12121."

Game 89

It is impossible, which can be proved by the rule of 9. (See solution to Game 68.)

The number to be squared has a digit sum of 1, 2, 3, 4, 5, 6, 7, 8, or 0 (representing 9). The square will have a digit sum corresponding to the squares of the digit sums:

$$1, 4, 9, 16, 25, 36, 49, 64, 0$$

Using the rule of 9 these digit sums become:

$$1, 4, 0, 7, 7, 0, 4, 1, 0$$

But the sum of 2, 4, 6, and 8 is 20, or 2 using the rule of 9, which can never correspond to a square.

Game 90

77	1	2	3	4	72	71	70	69
76	62	17	18	19	58	57	56	6
75	61	51	29	30	48	47	21	7
74	60	50	44	37	42	32	22	8
9	23	33	39	41	43	49	59	73
14	27	36	40	45	38	46	55	68
15	28	35	53	52	34	31	54	67
16	26	65	64	63	24	25	20	66
13	81	80	79	78	10	11	12	5

Game 91

The weights that balance on pans with unequal arms are not equal, but they are proportional: if 1 kilogram balances 8 melons, 2 kilograms will balance 16 melons, and so on.

Let P be the weight of a melon. We have:

$$\frac{1}{8P} = \frac{2P}{1}$$

Then:

$$16P^2 = 1$$
$$4P = 1$$
$$P = 250 \text{ grams}$$

Game 92

Only five pieces are necessary. Timothy divides the bar in parts of 1, 3, 9, 27, and 81 grams. In fact:

$$1 + 3 + 9 + 27 + 81 = 121$$

With this system, he can weigh any whole number of grams from 1 through 121 by:
- either balancing the desired weight with his pieces;
- or balancing the desired weight plus some of his pieces with other pieces.

For example, to weigh 20 grams, he puts 1 gram and 9 grams in one pan with the object, and 3 grams and 27 grams in the other pan.

Game 93

We will count the words by the placement of their F. All the F's form a diamond:
- in a corner of the diamond: an F touches 4 I's:
    1 I touches 7 ED's
    2 I's each touch 4 ED's
    1 I touches 1 ED

Thus each of 4 corner F's begins 16 FIED's and ends 16 DEIF's, which makes a total of $4 \times 16 \times 16 = 1024$ words
- on a side of the diamond: an F touches 4 I's:
    2 I's each touch 4 ED's
    1 I touches 2 ED's
    1 I touches 1 ED

Thus each of 8 side F's begins 11 FIED's and ends 11 DEIF's, which makes a total of $8 \times 11 \times 11 = 968$ words.

In all there are 1992 words.

Game 94

Let N be the number of $1 bills and P the number of $10 bills. We have:

$$1500 = N + 50N + 10P + 100P$$

$$1500 = 51N + 110P$$

Since 1500 and 110P are divisible by 10, N is divisible by 10 and we can write:

$$N = 10A$$

Dividing by 10, we have:

$$150 = 51A + 11P$$

Since 150 and 51A are divisible by 3, P is too, and we can write:

$$P = 3B$$

Dividing by 3, we have:

$$50 = 17A + 11B$$

The solution in positive integers is obvious:

$$A = 1, B = 3$$

$$N = 10, P = 9$$

Thus the cashier paid out:
- 10 bills of $1;
- 100 bills of $5;
- 9 bills of $10;
- 18 bills of $50.

Game 95

On each line the leftmost digit must be 1; otherwise it would not have been given.

The second intermediate product is two spaces to the left of the first instead of one. Therefore, the multiplier has a 0 in the middle only, and is 101:

$$
\begin{array}{r}
1\ *\ *\ * \\
1\ 0\ 1 \\
\hline
1\ *\ *\ * \\
1\ *\ *\ * \\
\hline
1\ *\ *\ 1\ *\ *\ *
\end{array}
$$

Since only 1 can be "carried" from the sum of two digits from intermediate products, the second digit of the product must be 0. To produce the 1 to be carried, the third digit of the product must also be 0, with a 1 above it; in turn, the fourth digit of the product (given as 1) must have a 1 above it so that 1 will be carried. The answer is:

$$
\begin{array}{r}
1\ 1\ 1\ 1 \\
1\ 0\ 1 \\
\hline
1\ 1\ 1\ 1 \\
1\ 1\ 1\ 1 \\
\hline
1\ 0\ 0\ 1\ 0\ 1\ 1
\end{array}
$$

In the decimal system: $15 \times 5 = 75$.

Game 96

Let X be the number of stations before the new line opened and Y the number of new stations.

Each new station yields $(2X + Y - 1)$ new tickets:

- tickets to every old station: X;
- tickets from every old station: X;
- tickets to every new station: $Y - 1$.

The total number of new tickets is:

$$Y(2X + Y - 1) = 34$$

The two numbers on the left are integers that divide 34. Then Y can only be 1, 2, 17, or 34. But 1 is impossible, for there are "several new stations." 17 and 34 are impossible too, for X would be negative.

There is only one solution:

$$Y = 2$$
$$X = 8$$

There were 8 stations. There are 2 new ones.

Game 97

Let X be the number of guests whose official language is French. Then there are 2X guests.

Every French-speaking guest says "Bonjour" to every other one and to the ambassador, that is, to X persons. This yields X^2 "bonjours."

Every non-French-speaking guest says "Bonjour" to every French-speaking guest and to the ambassador, that is to $(X + 1)$ persons, which yields $(X^2 + X)$ "bonjours."

Hence:

$$2X^2 + X = 78$$

Only the positive solution of the quadratic equation is correct:

$$X = 6$$

There were 12 guests.

Game 98

Every number begins with a 1.

From the skip in the second intermediary line, the quotient must be 1011. (Its second digit, which is skipped, is the only 0.)

The dividend and quotient end with 1, so the divisor does too.

If the hundreds digit of the divisor is 1, the second subtraction cannot give a three-digit result. So the divisor's hundreds digit is 0 and is below a 1 in the second subtraction.

If the tens digit of the divisor is 1, the second subtraction gives 101, so that the last subtraction will come out even as 1011 − 1011 = 0. But 1011 + 101 = 10,000, while the third line has only four digits. Therefore, the third digit of the divisor must be 0.

We have:

```
                 1 0 1 1
         1 0 0 1 | 1 1 0 0 0 1 1
                 1 0 0 1
                   1 1 0 1
                   1 0 0 1
                     1 0 0 1
                     1 0 0 1
```

Game 99

Internal doors are doors leading to other rooms, while external doors lead out of the house. Each room is either O (odd number of internal doors) or E (even number).

Each E room has an even number of external doors, since even (total doors) − even (internal) = even (external). The E rooms have a total number of external doors that is even, since even + even + . . . + even = even.

There is either an odd or an even number of O rooms. If it is odd, the O rooms have a total number of internal doors that is odd, since the sum of an odd number of odd numbers is odd. But every internal door opens on two rooms, and this would leave one door with no second room to open on.

Therefore, the number of O rooms is even. Each O room has an odd number of external doors to make its total number of doors even. The O rooms have a total number of external doors that is even, since the sum of an even number of odd numbers is an even number.

Since even (E rooms) + even (O rooms) = even, the whole house has an even number of external doors.

Game 100

1	92	8	94	95	96	97	3	9	10
20	12	13	84	85	86	87	88	19	11
71	29	23	74	75	76	77	28	22	30
40	39	38	67	66	65	64	33	62	31
50	49	48	57	56	55	54	43	42	51
60	59	58	47	46	45	44	53	52	41
70	69	68	37	36	35	34	63	32	61
21	72	73	24	25	26	27	78	79	80
81	82	83	17	15	16	14	18	89	90
91	2	93	4	6	5	7	98	99	100